Colorful

Dona Herweck Rice

Publishing Credits

Rachelle Cracchiolo, M.S.Ed., *Publisher*
Conni Medina, M.A.Ed., *Managing Editor*
Nika Fabienke, Ed.D., *Content Director*
Véronique Bos, *Creative Director*
Shaun N. Bernadou, *Art Director*
Carol Huey-Gatewood, M.A.Ed., *Editor*
Valerie Morales, *Associate Editor*
Courtney Roberson, *Senior Graphic Designer*

Image Credits: All images from iStock and/or Shutterstock.

Teacher Created Materials
5301 Oceanus Drive
Huntington Beach, CA 92649-1030
www.tcmpub.com

ISBN 978-1-4938-9816-9

Printed in Malaysia. Thumbprints.24790

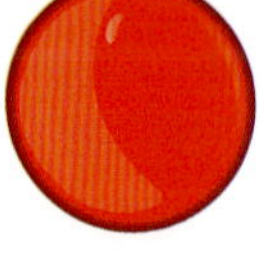

red

as a

crab

 as a 

orange pumpkin

 as a 

yellow duck

 as a 

white chicken

blue

as a

butterfly

 as a 

green leaf

 as a 

pink rose

 as a 

purple plum

● as a 

black crow

 as a 

gray whale

High-Frequency Words

a

as